THE WORLD OF OCEAN ANIMALS
PORTUGUESE MAN-OF-WARS

by Bizzy Harris

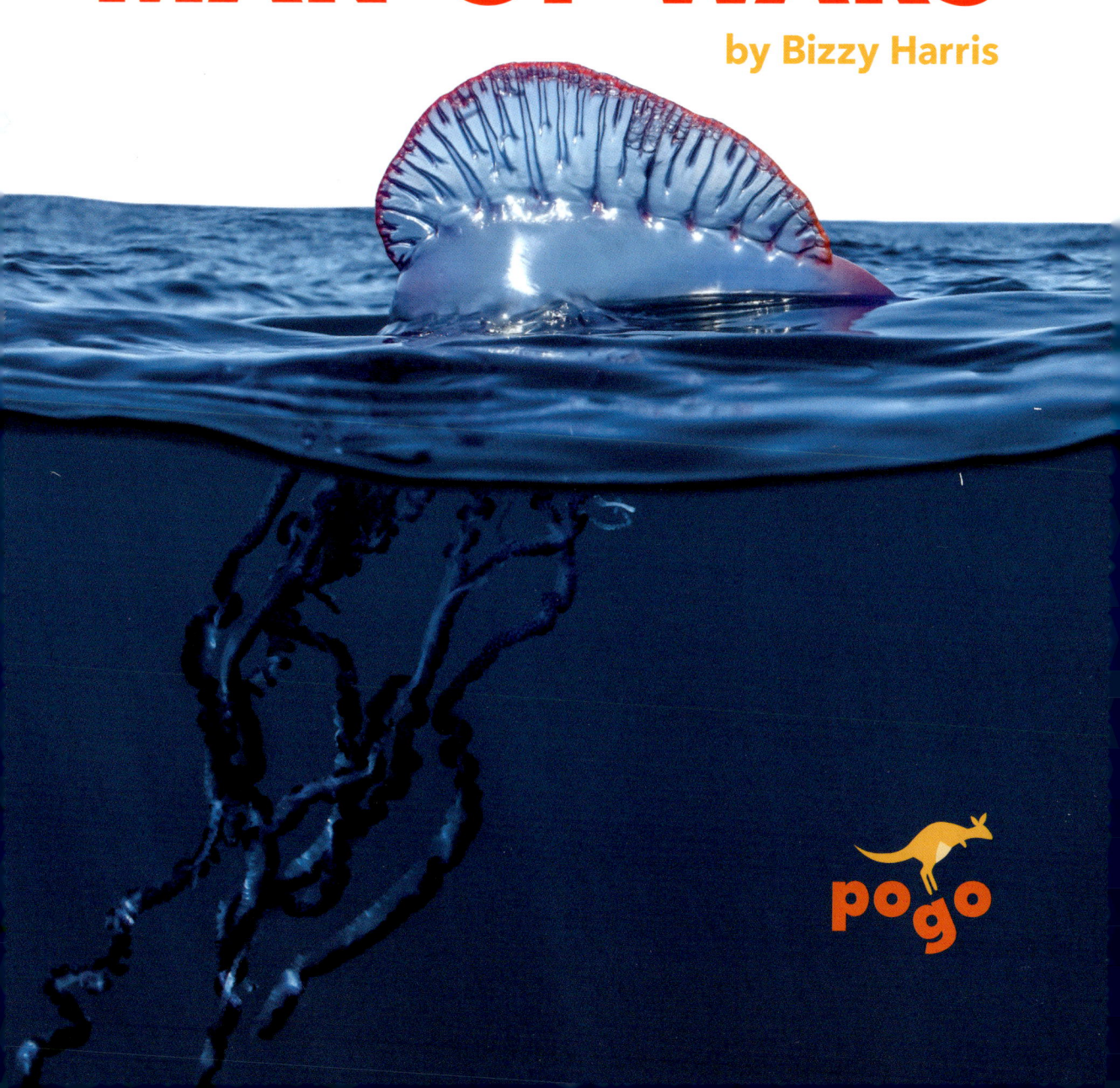

pogo

Ideas for Parents and Teachers

Pogo Books let children practice reading informational text while introducing them to nonfiction features such as headings, labels, sidebars, maps, and diagrams, as well as a table of contents, glossary, and index.

Carefully leveled text with a strong photo match offers early fluent readers the support they need to succeed.

Before Reading

- "Walk" through the book and point out the various nonfiction features. Ask the student what purpose each feature serves.
- Look at the glossary together. Read and discuss the words.

Read the Book

- Have the child read the book independently.
- Invite him or her to list questions that arise from reading.

After Reading

- Discuss the child's questions. Talk about how he or she might find answers to those questions.
- Prompt the child to think more. Ask: Did you know about Portuguese man-of-wars before reading this book? What more would you like to learn about them?

Pogo Books are published by Jump!
5357 Penn Avenue South
Minneapolis, MN 55419
www.jumplibrary.com

Copyright © 2022 Jump!
International copyright reserved in all countries. No part of this book may be reproduced in any form without written permission from the publisher.

Library of Congress Cataloging-in-Publication Data

Names: Harris, Bizzy, author.
Title: Portuguese man-of-wars / by Bizzy Harris.
Description: Minneapolis: Jump!, Inc., 2022.
Series: The world of ocean animals
Includes index. | Audience: Ages 7-10
Identifiers: LCCN 2021022459 (print)
LCCN 2021022460 (ebook)
ISBN 9781636902883 (hardcover)
ISBN 9781636902890 (paperback)
ISBN 9781636902906 (ebook)
Subjects: LCSH: Portuguese man-of-war–Juvenile literature.
Classification: LCC QL377.H9 H28 2022 (print)
LCC QL377.H9 (ebook) | DDC 593.5/5–dc23
LC record available at https://lccn.loc.gov/2021022459
LC ebook record available at https://lccn.loc.gov/2021022460

Editor: Jenna Gleisner
Designer: Michelle Sonnek

Photo Credits: IrinaK/Shutterstock, cover; by wildestanimal/Getty, 1, 8-9; Hailshadow/iStock, 3; Gonzalo Jara/Shutterstock, 4; Pally/Alamy, 5, 20-21; JaZa/Shutterstock, 6-7t; Humberto Ramirez/Getty, 6-7b; Reinhard Dirscherl/Alamy, 10, 23; Searsie/iStock, 11; IVANNE/Shutterstock, 12-13; Gerard Soury/Getty, 14; Grant Heilman Photography/Alamy, 15; Blue Planet Archive/Alamy, 16-17tl; Sahara Frost/Shutterstock, 16-17tr; Luiz Felipe V. Puntel/Shutterstock, 16-17bl; Richard Herrmann/Minden Pictures/SuperStock, 16-17br; Jim Simmen/Getty, 18-19.

Printed in the United States of America at Corporate Graphics in North Mankato, Minnesota.

TABLE OF CONTENTS

CHAPTER 1
A Floating Terror...4

CHAPTER 2
Surviving in the Ocean..................................10

CHAPTER 3
Prey and Predators.......................................14

ACTIVITIES & TOOLS
Try This!..22
Glossary..23
Index..24
To Learn More..24

CHAPTER 1
A FLOATING TERROR

What is that floating in the ocean? It is a Portuguese man-of-war!

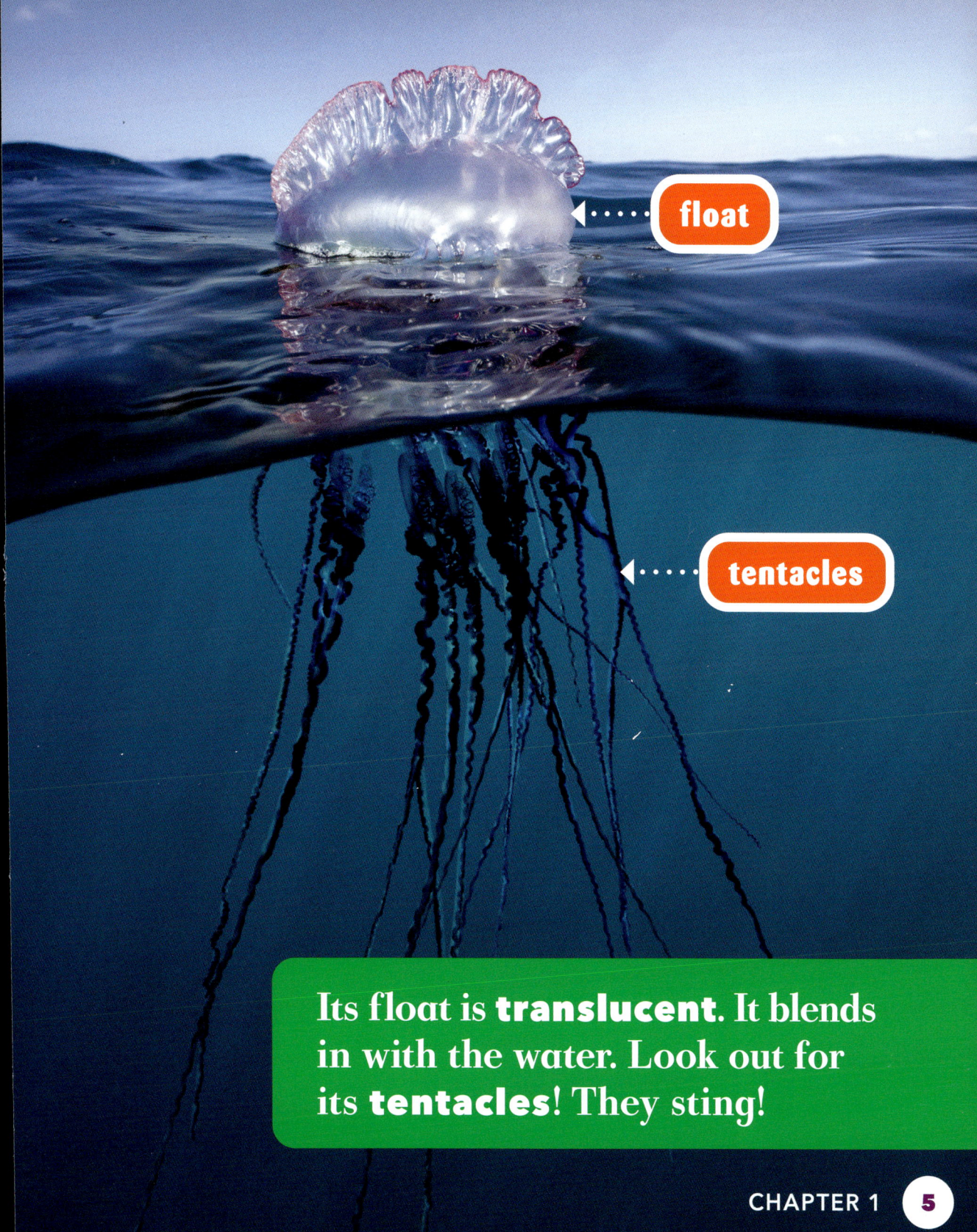

float

tentacles

Its float is **translucent**. It blends in with the water. Look out for its **tentacles**! They sting!

CHAPTER 1 5

jellyfish

man-of-war

CHAPTER 1

Man-of-wars are **invertebrates**. They don't have any bones. They look like jellyfish.

Like jellyfish, they don't have brains, hearts, or eyes. But unlike jellyfish, each man-of-war is a group of four **organisms**. These are called **zooids**. They work together as one.

DID YOU KNOW?

Man-of-wars are named after ships. Why? Their floats look like old Portuguese warships. They are also called floating terrors. Another name for them is bluebottles.

Each zooid has a job. The float is full of gas. It floats on the surface and helps a man-of-war move. Another zooid is for **reproduction**. The third **digests** food.

The tentacles are the fourth zooid. Each one has thousands of stinging **cells**. Most tentacles are about 30 feet (9.1 meters) long. But some can be 165 feet (50 m) long!

TAKE A LOOK!

Four zooids make up a Portuguese man-of-war. Take a look!

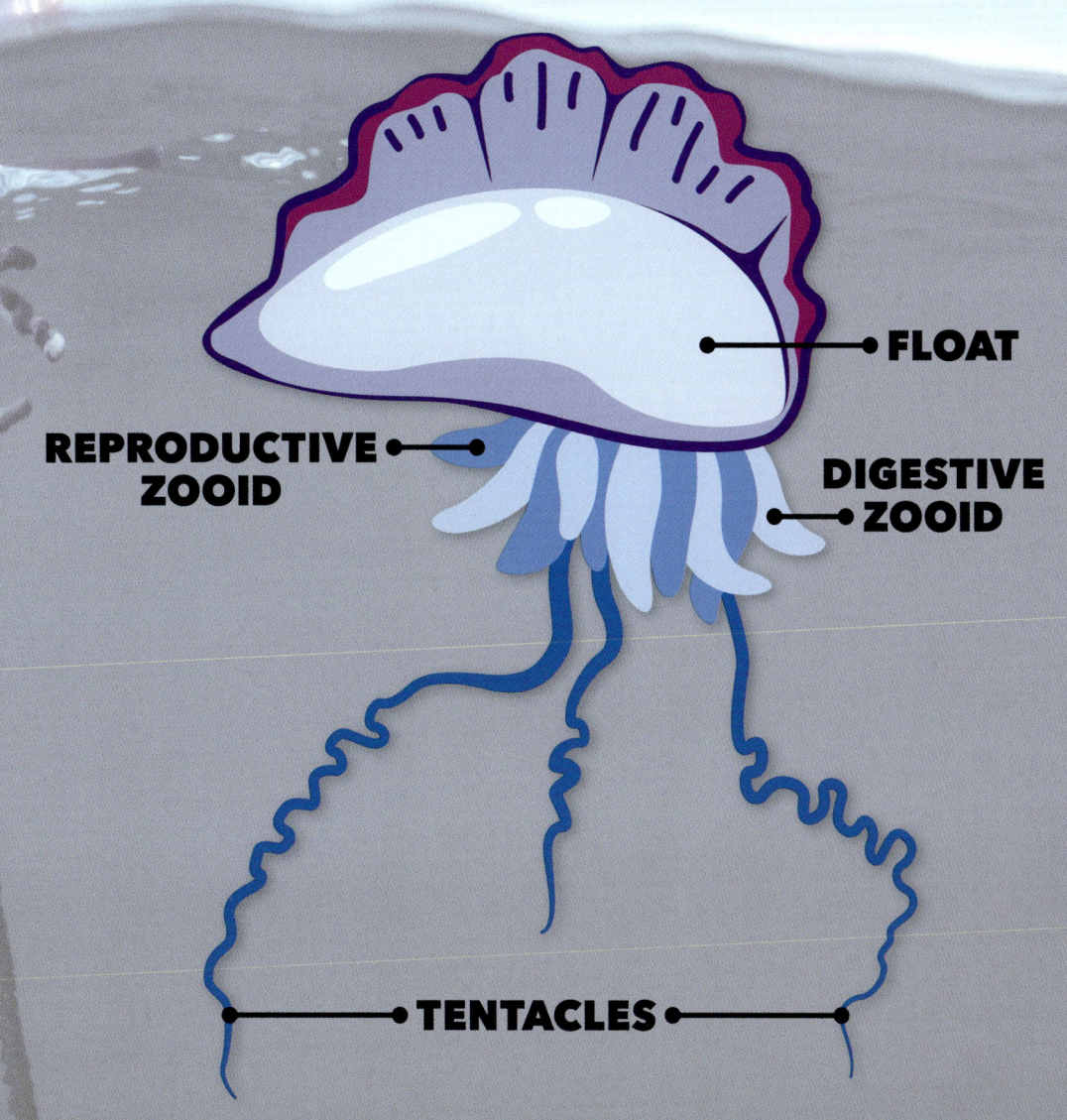

CHAPTER 1   9

CHAPTER 2

SURVIVING IN THE OCEAN

Man-of-wars do not swim. They drift on top of the water. Their floats catch the wind. Water **currents** move them, too.

Wind often pushes them together. Sometimes groups get pushed close to shore. They may even wash up on the beach.

CHAPTER 2　11

Man-of-wars live in warm ocean waters. They are known to live in the Atlantic, Indian, and Pacific Oceans. But their **range** could be even bigger! They live far out at sea, so it is hard for scientists to know.

DID YOU KNOW?

Floats on man-of-wars point in different directions. This helps them spread out. How? The wind pushes them in different directions.

CHAPTER 2

CHAPTER 3
PREY AND PREDATORS

Portuguese man-of-wars are **carnivores**. They mainly eat small fish. They can eat up to 120 each day! Fish get caught in their tentacles.

The tentacles sting and **paralyze** the fish. It can't swim away. The tentacles then move the fish to the digestive zooid. The zooid breaks down the food.

- blanket octopus
- blue dragon sea slug
- loggerhead sea turtle
- ocean sunfish

The man-of-war has some unique **predators**. The blanket octopus is one. After it feeds on a man-of-war, it uses its tentacles. How? They attach to its suckers. The octopus uses them to fight predators.

The blue dragon sea slug also uses the man-of-war's tentacles. It stores the stinging cells in its body. Loggerhead sea turtles and ocean sunfish also eat man-of-wars.

CHAPTER 3

Sometimes man-of-wars **deflate** their floats. Why? They sink below the surface. This helps them hide from predators.

DID YOU KNOW?

Man-of-wars can **sense** what is around them. How? Special cells are in the tentacles and around the digestive zooid. They sense temperature changes and touch.

CHAPTER 3

man-of-war fish

Man-of-wars protect man-of-war fish. These fish live and swim among a man-of-war's tentacles. Predators can't get them. They avoid the tentacles as they swim. They don't get stung!

The man-of-war is an interesting sea creature. What else would you like to learn about it?

ACTIVITIES & TOOLS

TRY THIS!

WORK TOGETHER AS ONE

Learn how to work with others, like the zooids that make up a man-of-war, in this fun activity!

What You Need:
- three friends or classmates
- red, blue, and green construction paper
- scissors
- a clock or watch

❶ Cut five circles from each color of construction paper. Green will be food, blue will be man-of-wars, and red will be predators.

❷ Place the circles on the floor around the room.

❸ Assign each person on your team a task. You and your team are the zooids that make up a man-of-war! The float will decide where to go. The reproductive zooid will find other man-of-wars. The digestive zooid will find food. The tentacles will sting predators.

❹ Link arms with your teammates so you are all facing outward in a circle. The room is your ocean.

❺ Set a timer for two minutes. When the timer starts, see if you and your group can find food and other man-of-wars while keeping the predators away. Pick up the green and blue circles. Stay away from the red circles!

❻ Discuss whether it was difficult to work together as one. How do you think this is the same or different from how real zooids work together?

GLOSSARY

carnivores: Animals that eat meat.

cells: The smallest units of animals or plants.

currents: Movements of water in one direction.

deflate: To let the air out of something.

digests: Breaks down food so that it can be absorbed and used by the body.

invertebrates: Animals that do not have backbones.

organisms: Living things, such as plants or animals.

paralyze: To make someone or something unable to function.

predators: Animals that hunt other animals for food.

range: The area where an animal lives.

reproduction: The act of producing offspring or individuals of the same kind.

sense: To feel or become aware of something.

tentacles: The flexible limbs of some ocean animals.

translucent: Clear enough to allow light through.

zooids: Individual living things that make up compound organisms.

INDEX

beach 11
bones 7
brains 7
currents 10
deflate 18
eat 14, 17
eyes 7
float 5, 7, 8, 9, 10, 12, 18
group 7, 11
hearts 7
jellyfish 7
man-of-war fish 21
move 8, 10, 15
named 7
ocean 4, 12
organisms 7
predators 17, 18, 21
scientists 12
shore 11
sting 5, 15
stinging cells 8, 17
tentacles 5, 8, 9, 14, 15, 17, 18, 21
wind 10, 11, 12
zooids 7, 8, 9, 15, 18

TO LEARN MORE

Finding more information is as easy as 1, 2, 3.

❶ Go to www.factsurfer.com
❷ Enter "Portugueseman-of-wars" into the search box.
❸ Choose your book to see a list of websites.